REPORT

OF

JOHN A. ROEBLING,

CIVIL ENGINEER,

TO THE

DIRECTORS OF THE NIAGARA FALLS INTERNATIONAL

AND SUSPENSION BRIDGE COMPANIES.

BUFFALO:

STEAM PRESS OF JEWETT, THOMAS & CO.

1852.

REPORT.

Your bridge will form a single span of 800 feet, from center to center of towers. It is to serve as a connecting link between the rail roads of Canada and those of the State of New York; and is also to accommodate the common travel of the two countries.

The question of the practicability of a suspension rail road bridge has been variously discussed. Some of the most eminent engineers have decided against it. Mr. Robert Stephenson, among others, made a report upon the failure of a rail road suspension bridge in England, and used it as an argument against the application of suspension chains for crossing the Menai Straits. In this report, this eminent engineer does no justice to the question; his only aim appeared to be to remove the objections, which were raised by the advocates of the suspension principle, against his own plan of a tubular bridge.

Whenever necessity calls for new works of art, new expedients will be discovered, adapted to the occasion. It will no longer suit the spirit of the present age to pronounce an undertaking impracticable. Nothing is impracticable, which is within the scope of natural laws.

The principles which have guided me in the planning of your work, are exceedingly simple, and do not involve any intricate question. The only real difficulty of the task appears to be its novelty. Two principal considerations present themselves: that of strength and stiffness. In regard to strength, it is not only admitted by all parties, but established by ample experience, that good iron wire, if properly united into cables or ropes, is the best material for the support of loads and concussions, in virtue as well of its great absolute cohesion, which amounts to from 90,000 to 130,000 lbs. per superficial inch, according to quality and fineness, as on account of its high degree of elasticity, which in good, hard-drawn wire, approaches that of steel. To support a heavy weight, nothing safer can be employed, than a well-made wire cable or rope. It is not only the strongest, but also the most economical. The second question is that of stiffness. Wire cables, if large, well made, and compactly wrapped, possess a considerable degree of stiffness, but too little to be taken into account in the present case. We can depend upon cables only for support, and must apply other means for producing stiffness. There are three different modes for obtaining stiffness:

1. By using girders and heavy timbers, longitudinally, in the construction of the floor.

2. By trusses.

3. By stays.

The plan of the double floor bridge, which you have adopted, provides two girders in the upper floor, of a depth of 4 feet for the immediate support of the rail road track. This is a very important feature in the construction of any rail road bridge; it serves to distribute the pressure upon any one point over a greater length, and thereby prevents short impressions and undulations. Such girders would alone be sufficient for moderate spans, but larger ones require trusses besides. Since you have decided in favor of a double floor, the upper one to be used for the rail road, the lower one for common travel, a very good opportunity has offered for doubling the trusses, and of adding another valuable feature, that of the box or tube. Your bridge will form a hollow, straight beam, of 20 feet wide and 18 feet deep, composed of top, bottom and sides. The upper floor, which supports the rail road, is 24 feet wide between the railings, and suspended to two wire cables, assisted by stays. The lower floor is 19 feet wide in the clear, connected with the upper one by vertical trusses, forming its sides, and suspended to two other cables, which have 10 feet more deflection than the upper cables. The mechanical principles upon which the trusses I have devised for this bridge, act, is not new, but universally applied. On inspection of the plan, you will observe posts, in a vertical position, 5 feet apart, firmly connected with the upper and lower floor beams. Any pressure upon the upper floor will, by these posts, be transferred upon the lower, and vice versa. With the exception of these, and a light bridging between, there is no other timber to be used in this truss. Its stiffness is to be obtained from wrought iron rods, of one inch diameter, and 27 feet 4 inches long, which are placed diagonally, nearly at right angles to each other, and always connecting the upper and lower end of the first and fifth post. The ends of these rods have screws and nuts, and are screwed up as tight as the strength of the posts will permit. The result of this is a high degree of rigidity and stiffness. The vertical action of any one post is by these tension rods transmitted to two other posts, 40 feet apart.

This plan recommends itself not only on account of its great stiffness, but also on account of its great lightness, and freedom from all timber-joints and connections. Consequently no working or opening of joints can take place. The manner in which the posts are framed and connected with the upper and lower floor, and the fact, that the rods are only exposed to tension, insure the true vertical plane of the trusses; they are not exposed to the danger of drawing out of line.

All the timber used in this structure, is to be well seasoned, planed, and well painted. The upper floor will be treated like a ship's deck, caulked and painted, it will therefore serve as a complete roof for the lower work.

The trusses are so open, that they present but little surface to the action of the wind.

The upper floor stays will be a great additional support to the upper cables, and will add materially to the stiffness of the structure. These and the lower floor stays, which I propose to put up, will in connection with the trusses and the girders, produce a degree of stiffness, sufficient, I feel confident, to admit of the passage of the heaviest trains at full speed, without producing any injurious vibrations. The ordinary speed of trains, however, is to be limited to five miles per hour.

THE ANCHORAGE

Will be formed by sinking 8 shafts into the rock 25 feet deep. The bottom of each shaft will be enlarged for the reception of cast iron anchor plates of 6 feet square. These chambers will have a prismatical section, which when filled with solid masonry can not be drawn up without lifting the whole rock to a considerable extent. This resistance is further increased by the weight of the superincumbent masonry. My calculations satisfy me, that the pressure upon the anchor plates will be met with a resistance equal to the ultimate strength of the cables, or equal to five times the maximum tension to which they can ever be exposed. To these anchors massive iron chains are attached, which rise vertically through the rock, form a curve in the anchor masonry, and connect with the wire cables at the level of the coping. Each chain will be 66 feet long and composed of 8 links, each link to be formed of 7 or 8 plates of an aggregate section of 72 superficial inches of solid wrought iron. The links connect by means of bolts, turned off to $3\frac{1}{2}$ inch diameter, and passing through eyes accurately bored out.

The plates will be forged out of the best quality of charcoal blooms, and in one piece without a weld. By the curvature of the chain, a part of its tension is transformed into pressure upon the wall, and will be supported by large cut stone blocks. The chains and pins will be well painted, and embedded in grout, composed of cement, lime, and sand.

SADDLES

Of cast iron will support the cables on top of the towers. They will consist of two parts, the lower one stationary, the upper one movable, resting upon wrought iron rollers of 3 inches diameter, accurately turned off, and fitting against the faces of the upper and lower plate, which are to be planed off true. The rollers will admit of a slight motion of the upper saddle, whenever the balance of tension between the land and suspension cables should be greatly disturbed. The saddles will have to support a pressure of 600 tons whenever the bridge is loaded with a train of a maximum weight.

TOWERS.

They are 60 feet high, 15 feet square at the base and 8 feet square at the top; therefore the top course, on which the saddle rests, has a surface of 64 square feet. The compact, hard limestone, which will be used in the masonry of the towers, will bear a pressure of 500 tons upon every foot square. The strength of the towers to support the above weights, can therefore be depended upon. As regards their lateral stability, I rely solely upon their own weight and the great pressure they have to support, which pressure will ordinarily act in a vertical direction through the center of the masonry.

WEIGHT OF BRIDGE.

Weight of Timber,	919,130	lbs.
" Wrought iron and Suspenders,	113,120	"
" Castings,	44,332	"
" Rails,	66,740	"
" Cables between Towers,	535,400	"
Total weight of Superstructure,	1,678,722	"

As the ends of the bridge rest upon the rock and are firmly connected with the masonry of the abutments, the trusses will support them for *at least* 40 feet, independent of the cables, and together with the loads. Suppose the ends were not suspended to the cables for 100 feet from the center of tower toward the center of the span, but only supported by the abutment at one end and at the other by the cables, then, I say, there would be strength enough in the trusses to support themselves and loads. The length of bridge being 800 feet from center to center of abutment, there are only 700 feet actually supported by the cables, and 100 feet by the abutments.

Reducing the above weight proportionally, it leaves the weight of superstructure, as far as supported by the cables, 1,564,391 lbs., or 782 tons of 2000 lbs. each.

WEIGHT OF RAIL ROAD TRAINS.

One Locomotive of 30 feet length weighs, say,	25	tons.
27 Double Freight Cars, each 25 feet long, and of 15 tons gross weight,	405	"
Make total gross weight,	430	tons,

which will fall upon the cables, when the whole bridge is covered by a train of cars from end to end.

Add to this 15 per cent. increase of pressure, as the result of a speed of five miles per hour, which is a very large allowance, or	61	tons,
Makes total,	491	tons.
Add weight of Superstructure,	782	"
Total aggregate maximum weight,	1273	tons.

Since the grading of the two rail roads will scarcely permit trains to be run of the weight and magnitude above assumed, we are certain that the above maximum can never be exceeded.

TRANSIENT LOADS.

The area of the lower floor, as far as supported by the cables, is, ... 700×19=13,300 superficial feet.

Supposing the floor covered with a crowd of people, and allowing 3 superficial feet for each person,

There could be placed upon the floor, ... 4433 men,

Which, at 140 lbs. would weigh, ... 620,620 lbs.

Or, ... 310 tons.

There is, however, no necessity to calculate the strength of the bridge for maximum loads upon both floors. In case an unusual crowd of persons should at any time collect or cross upon the lower floor, there can be no necessity for passing heavy trains over the upper floor at the same time. They can be stopped, until the lower floor is cleared, and vice versa. With gates at the two ends of the upper floor, this rule can always be enforced.

CABLES.

There will be four wire cables of $9\frac{1}{4}$ inch diameter each. Two will be suspended with a deflection of 54 feet for the support of the upper floor. The other two will extend down the sides of the trusses, for the support of the lower floor, with a deflection of 64 feet. Average deflection is 59 feet. When a train of cars enters upon the bridge, the upper cables will at once be taxed, but the pressure will, at the same time, be transferred upon the lower ones, so that they will always act in concert. The reason why I do not deflect the outside cables still lower, is because their harmonious action would thereby be disturbed, on account of unequal contraction and expansion, which would take place either in consequence of changes of temperature, or of the depressing action of great transitory weights. The calculations I have made to that effect, have satisfied me, that I may extend the difference in the deflection to 10 feet safely, but no further.

The tension of the cables, which would result from a weight of 1273 tons, and an average deflection of 59 feet, is

$$1273 \times 1.76 = 2240 \text{ tons.}$$

Since the above assumed maximum tension can but rarely occur, I consider it ample to allow four times the strength to meet it; or,

$$4 \times 2240 = 8960 \text{ tons.}$$

But assuming 2000 tons as a tension to which the cables may be subjected more frequently, I propose to allow five times the strength to meet it, or to provide for an ultimate strength of

10,000 tons.

The Act of Parliament calls for an ultimate strength of 6,600 tons, for which your former engineer allowed 10,000 wires of No. 10. Now, the best charcoal wire, drawn to No. 10, or measuring $20\frac{1}{2}$ feet for every pound, cannot, upon an average, be depended upon for more than 1333 lbs.; or at the rate of $1\frac{1}{2}$ wires per ton of 2000 lbs. Being myself engaged in the manufacture of wire, and of wire rope, and having ample experience in the construction of wire cables, the above rate of strength, as the result of my own experience and tests, may be depended upon as correct. The usual mode of testing wire by weights is not to be relied upon. Common wire, made of puddled anthracite iron, will frequently exhibit the same strength as wire made of good cold blast charcoal iron.

But the difference is in the uniformity, and this can only be examined by suspending a long wire say 400 feet, freely between posts, and reducing its deflection by gradually hauling in the one end by means of a capstan, until it breaks; the tension being the result of its own weight. By noting the deflection, the strength of the wire may then be calculated in pounds for any size or per superficial inch. Thus tested, the wire of course will break at the weakest point, and common puddled wire will thus show its inferiority to good charcoal wire.

Assuming an ultimate strength for the cables and stays of 10,000 tons, we shall require 15,000 wires of No. 10 to construct them.

At each end of the upper floor, the upper cables will be assisted by 18 wire-rope stays, and their strength will be equivalent to 1440 wires, these deducted, leave the number of wires in the four suspension cables, 13,560 ; number of wires of one cable, 3,390 ; diameter of cables, $9\frac{1}{4}$ inch.

REMARKS ON A SINGLE FLOOR BRIDGE.

My original estimate for a single floor bridge was made out for a floor of 30 feet wide between the railings, viz :

For Railroad track in center, ...	6 feet.
" Two tracks for common travel @ 7 feet, . . .	14 "
" Two footwalks, @ 4 feet, ...	8 "
" Space occupied by cables, ...	2 "
Total between railing, ...	30 "

On such a floor the common carriage travel could not be passed, while the track is occupied by a train, and vice versa, it would therefore only accommodate a limited business.

To render the rail road independent of the common travel, we should want a width of floor of no less than 39 feet between the railings, viz :

For railroad in center, ...	13 feet
" 2 tracks for common travel, @ 8 feet, ...	16 "
" 2 sidewalks, ...	8 "
" 2 cables, ...	2 "
Total, ...	39 "

The entrance upon this floor between the towers would have to be at least 32 feet, and in consequence an arch would be necessary, to connect them, for the support of the center cables. The two outside cables would fall upon the shafts of the towers. We should want gateways of large dimensions, no less than 65 feet by 30 at the base, with abutments in proportion, and a mass of masonry three times as great as what is wanted for the double floor bridge. To preserve the character of the work, the wings would have to be carried up to the abutments. The great width of floor, would require heavier crossbeams, these with trusses underneath, which could not be dispensed with, would rather increase than diminish the weight of the bridge. All this satisfies me, that a single floor bridge of the above dimensions would cost considerably more than the double floor bridge.

The appearance of the double floor bridge will not be so beautiful, as that of a single floor, with imposing gateways, erected in the massive Egyptian style, and joined by massive wings, the cables watched by sphinxes, with parapets, and all the rest of the approaches, put up of appropriate dimensions and in suitable style. The double floor bridge, when viewed from the upper floor, where all the foot passengers will resort, will, however, present a very graceful, simple, but at the same time substantial appearance. The four massive cables, supported on isolated columns of a very substantial make, will form the characteristic of the work, and this will be unique and striking in its effect, and quite in keeping with the surrounding scenery.

Niagara Falls, July 28th, 1852.

www.ingramcontent.com/pod-product-compliance
Lightning Source LLC
LaVergne TN
LVHW020642110826
845149LV00004B/1319

9781418189679